FLORA OF TROPICAL EAST AFRICA

ONAGRACEAE

J. P. M. Brenan

Herbs, sometimes shrubs, very rarely trees. Leaves alternate, opposite or rarely whorled, simple ; petiole not inflated in middle. Stipules absent or small. Flowers solitary and axillary, sometimes aggregated into racemes or panicles, normally hermaphrodite, regular or rarely irregular. Sepals valvate, mostly 4, sometimes 5, rarely 2, 3, 6 or 7. Petals free, as many as sepals, contorted or imbricate, rarely absent. Stamens as many or twice as many as sepals, very rarely fewer. Ovary inferior, mostly 4-celled, rarely 1–7-celled. Style simple. Ovules axile, one to many per cell. Fruit a capsule, rarely a nut or a berry. Seeds without endosperm.

This family contains many ornamental plants likely to be met with in cultivation. In our area I have evidence of the following genera being grown : *Fuchsia* L., *Gaura* L. and *Oenothera* L. (subgenus *Hartmannia* (Spach) Munz).

Seeds with an apical tuft of silky hairs ; sepals caducous
    after flowering ; petals mostly pink to purple, some-
    times whitish . . . . 1. **Epilobium**
Seeds without any hairs :
  Receptacle not or scarcely prolonged beyond ovary ;
    sepals persistent after flowering ; fruit a capsule ;
    flowers (in our species) yellow . . . . 2. **Jussiaea**
  Receptacle prolonged beyond ovary into a distinct usually
    coloured " calyx-tube " or hypanthium which is
    caducous after flowering ; fruit a berry ; flowers (in
    our species) purple . . . . . . 3. **Fuchsia**

1. EPILOBIUM

L., Gen. Pl., ed. 5, 164 (1754)

Annual or perennial herbs, rarely becoming thinly woody. Leaves oppo-site, alternate or sometimes whorled, in our species denticulate or serrulate to subentire. Stipules 0. Flowers normally regular. Sepals 4, caducous after flowering. Receptacle or " calyx-tube " not or hardly prolonged beyond the ovary. Petals 4, mostly pink, red or purple, rarely white or yellow. Stamens 8, the four antepetalous ones often shorter than the others. Stigma clavate or capitate, entire or more or less 4-cleft. Ovules uniseriate in each cell of the 4-celled ovary. Fruit an elongate capsule splitting loculi-cidally into 4 valves. Seeds each with a conspicuous apical tuft of numerous silky normally whitish hairs.

In Europe the species of *Epilobium*, including *E. hirsutum* L., hybridize very freely, the resultant plants being characteristically sterile or nearly so. I have no convincing evidence for such hybrids in East Africa, but observers will do well to look out for them wherever two or more species grow together.

Plant (in Africa) more or less densely villous-hairy ;
    stigma four-lobed nearly or quite to base   .   .   1. *E. hirsutum*
Plant subglabrous, puberulous or at most pubescent ;
    stigma entire or very shortly lobed (lobes to 1/3
    stigma-length) :
Stems at extreme base with zone of imbricate scale-
    leaves ; short spreading glandular hairs present
    among pubescence (use × 20 lens) ; stigma
    capitate, wider than long, not tapering below   2. *E. sp.*
Stems without imbricate scale-leaves at base ;
    glandular hairs absent ; stigma usually longer
    than wide, sometimes (*E. stereophyllum*) as wide
    as long :
Cauline leaves normally broadly rounded to sub-
    cordate at base ; flowers pink or mauve, even
    when just open ; petals normally distinctly
    obovate, 4–11 mm. wide ; stigma obovoid-
    clavate, often nearly as wide at apex as long   3. *E. stereophyllum*
Cauline leaves cuneate to narrowly rounded at
    base ; flowers white or cream, often becoming
    pink after opening ; petals with elliptic or
    narrowly obovate lamina, usually 4·5 mm. or
    less wide ; stigma usually oblong to ellipsoid,
    distinctly longer than wide   .   .   4. *E. salignum*

1. **E. hirsutum** *L.*, Sp. Pl. 347 (1753), excl. var. β (=*E. parviflorum* (Schreb.)
Schreb.) ; Hausskn., Monogr. Epilobium 53 (1884). Type : presumably
from Europe, *Herb. Linnaeus* (LINN, lecto.!)

Rhizomatous, usually robust herb up to 2·1 m. high, indumentum ranging
from pubescent and very sparsely villous to densely villous or tomentose ;
capitate glandular hairs present. Stems without imbricate scale-leaves at
base. Leaves sessile, variable, oblong or lanceolate to elliptic or sometimes
ovate-elliptic or obovate-oblong, about 3–10 cm. long and 0·5–2·5 (–3·7) cm.
wide, acute, sharply serrulate. Flowers bright pink to red-purple, rarely
and only casually white. Sepals 7–10 mm. long. Petals broadly obovate,
9·5–19 mm. long, 7–18 mm. wide. Style 7–8 mm. long. Stigma divided
nearly or quite to base into four arcuate-recurved narrowly clavate obtuse
arms 1·75–3 mm. long. Capsules mostly 4–8·5 cm. long ; pedicels 0·5–1·6
(–2) cm. long.

var. **villosissimum** ("*villosissima*") *Koch*, Syn. Fl. Germ. Helv., ed. 1, 240 (1835).
No specimens or localities cited, but type-locality presumably central Europe

Plant more or less densely villous-hairy on leaves, stems, ovaries, sepals and capsules.
Leaves (in Africa) lanceolate, very acute, mostly 0·5–1·7 cm. wide. Petals (in Africa)
mostly up to about 11 mm. long and 9 mm. wide. Fig. 1/1.

UGANDA. Toro District : E. Ruwenzori, Bwamba Pass, July 1940, *Eggeling* 4015! ;
    Kigezi District : Kigezi, Lake Mutanda, Jan. 1947, *Purseglove* 2312!
KENYA. Elgon, Jan. 1931, *Major & Mrs. Lugard* 499! ; Masai District : Ngong,
    Napier 520!
TANGANYIKA. Masai District : Ngorongoro crater, Laroda, 13 Sept. 1932, *Burtt* 4337! ;
    Ufipa District : Ufipa Plateau, E. side of Kalambo Basin, 14 Dec. 1934, *Michelmore*
    1059! ; Kondoa District : Mondo, 1 Jan. 1928, *Burtt* 1022!
DISTR. U1, 2 ; K3–6 ; T2–5, 7 ; Europe from Scandinavia southwards, the Medi-
    terranean Region, extending through Asia to China, and through eastern Africa to
    Angola and the Cape
HAB. Swampy and marshy places by rivers, streams, ditches and lakes ; 1190–2590 m.

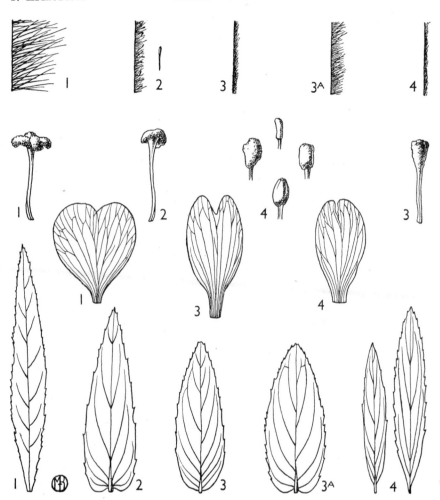

FIG. 1. *EPILOBIUM*—Top row : ovary indumentum, × 20 ; second row : stigmas, × 3 ; third row : petals, × 3 ; bottom row : leaves, natural size. 1, *E. hirsutum* var. *villosissimum;* 2, *E. sp.* No. 2 (page 4) ; 3, *E. stereophyllum* var. *stereophyllum;* 3A, *E. stereophyllum* var. *kiwuënse* ; 4, *E. salignum.*

SYN.   [*E. hirsutum* L. var. *intermedium* (Mérat) [Ser. ex] DC., Prod. 3 : 42 (1828) *quoad descr. tantum, non E. intermedium* Mérat, Fl. Paris 147 (1812) *quod, e descr. et fide Hausskn.,* Monogr. Epilobium 71 (1884), *est E. parviflorum* (Schreb.) Schreb.]
   E. hirsutum L. var. *villosum* Hausskn., Monogr. Epilobium 55 (1884). Many localities and specimens mentioned, but no type indicated
   E. hirsutum L. f. *valdepilosum* Saut. (" L.") in Oesterr. Bot. Zeitschr. 49 : 402 (1899). No type-specimen cited, but a locality : Tyrol, Salurn
   E. hirsutum L. var. *africanum* Léveillé in Bull. Herb. Boiss., ser. 2, 7 : 589 (1907). Type : " Africa australis," 1894, *Wilms* 464 (G, holo. !)

In Europe and Asia *E. hirsutum* and also var. *villosissimum* are very variable in such things as size and shape of leaves and size of flowers. In tropical and South Africa the reverse is true, var. *villosissimum* being on the whole unusually uniform through this part of its range. One might follow Léveillé in making the African plant a distinct variety, but I do not consider it sufficiently distinct from some forms of var. *villosissimum* in Europe. This position suggests the possibility of a relatively recent and rapid migration southwards of a single race capable of withstanding African conditions.

E. hirsutum is the familiar and lovely large-flowered willow-herb or Codlins and Cream so conspicuous over the English countryside in summer.

Distr. (of species as whole). Similar to that of var. *villosissimum*. *E. hirsutum*, in a wide sense, recorded as naturalized in Canada and the United States

2. E. sp.

Perennial herb up to about 1 m. high. inconspicuously pubescent with short curved simple hairs mixed with more or less numerous straighter spreading capitate glandular hairs (use × 20 lens). Stems subsimple or sometimes branched, at extreme base with a zone of densely imbricate thick scale-leaves, each one rounded and wider than long. Leaves mostly ovate or ovate-lanceolate, about 2·5–5 cm. long and 1–2 cm. wide, subacute or acute at apex, broadly rounded or subcordate at base ; petiole 1–2 (–3) mm. long. Flowers drooping, pink or crimson. Sepals 6–7·5 mm. long. Petals [described from poor and inadequate material] obovate, about 11·5–13 mm. long and 5 mm. or more wide. Style about 7·5 mm. long. Stigma not lobed, capitate, not tapering below, wider than high, about 1·5 mm. high and 2·5 mm. wide. Capsules about 5–7·5 cm. long ; pedicels 3·5–6·5 cm. long. Fig. 1/2, (p. 3).

Tanganyika. Njombe District : Madahani, 3 Dec. 1913, *Stolz* 2341 ! ; Rungwe District : S. slopes of Poroto Mts., Ndwati Lake, Manipindi's, 16 Mar. 1932, *St. Clair Thompson* 918 ! ; Iringa District : Mufindi, Kigogo Forest Reserve, 1 Feb. 1934, *Michelmore* 953 !
Distr. **T7** ; endemic to south-western Tanganyika
Hab. Upland grassland ; 1830–2100 m.

This species, so distinctive in stigma-shape, scale-leaves and glandular hairs, is not as closely related to the other species in our area as it is to the South African *E. flavescens* [E. Mey. ex] Harv. & Sond., which has yellowish flowers, a similarly shaped stigma but with four small but distinct lobes, and often shorter capsules and pedicels.
The remarkable scale-leaves at the base of the stem, which look as though they were fleshy when living, probably act as storage-organs. Botanists in south-western Tanganyika are asked to verify this, and also to collect more of this species, of which the herbarium material is still inadequate. This inadequacy makes me think it wiser to await better material before describing the species as new.

3. E. stereophyllum *Fres.* in Mus. Senckenb. 2 : 152 (1837) ; Hausskn., Monogr. Epilobium 233 (1884) ; Léveillé, Iconogr. Epilobium, t. 47 (1910). Type : Abyssinia, between Temben and Semen, *Rüppell* (FR, holo. !)

Stoloniferous herb to 1 m. high, often less, puberulous or shortly pubescent especially on midribs, ovaries and pedicels ; leaves and stems often appearing subglabrous ; no glandular or capitate hairs present. Stems without imbricate scale-leaves. Leaves ovate to ovate-lanceolate or oblong, about 1·5–5·5 cm. long and 0·3–2·1 (–2·6) cm. wide, obtuse or subacute at apex, the cauline ones normally broadly rounded to subcordate at base ; petiole almost absent, 0·5–2·5 mm. long. Flowers pink or mauve, even when just open. Sepals 5·5–10 mm. long. Petals mostly obovate, 8–15 mm. long, 4–11 mm. wide. Style 5-6·5 mm. long. Stigma obovoid-clavate, tapering below, often nearly as wide at apex as long, with 4 short lobes at apex less than a third of the total stigma-length. Capsules 3·5–7·7 cm. long ; pedicels (1–) 2–6·5 cm. long.

var. stereophyllum

Ovaries and capsules, often also young stems, midribs and margins of leaves, densely puberulous with more or less appressed hairs ; indumentum less than 0·25 mm. high. Fig. 1/3, (p. 3).

Uganda. Ruwenzori, Aug. 1938, *Purseglove* 238 ! ; Kigezi District : Muhavura-Mgahenga saddle, Sept. 1946, *Purseglove* 2146 ! ; Mbale District : Bulambuli, 4 Sept. 1932, *Thomas* 512 !

Kenya. Elgon, Swam Valley, 16 May 1948, *Mrs. Adamson* AJ. 491! ; Elgon, Swam River, 18 May 1948, *Hedberg* 1000! ; Mount Kenya, 19 Nov. 1933, *C. G. Rogers* 671! Tanganyika. Kilimanjaro, without date, *Thomson*! ; Kilimanjaro, Oct. 1894, *Volkens* 1847! ; Kilimanjaro, N. slopes, above Rongai, 1 Dec. 1932, *C. G. Rogers* 162! Distr. U2, 3 ; K3, 4 ; T2 ; Abyssinia

Syn. *E. fissipetalum* [Steud. ex] A. Rich., Tent. Fl. Abyss. 1 : 273 (1848) ; Hausskn., Monogr. Epilobium 234 (1884) ; Léveillé, Iconogr. Epilobium, t. 50 (1910). Type : Abyssinia, Mt. Buahit, *Schimper* 1348 (P, holo., K, iso.!)
E. cordifolium A. Rich., Tent. Fl. Abyss. 1 : 274, t. 50 (1848) ; Hausskn., Monogr. Epilobium 233 (1884) ; Léveillé, Iconogr. Epilobium, t. 48 (1910). Type : Abyssinia, *Quartin-Dillon & Petit* (P, holo.!)
E. kilimandscharensis [sic] Léveillé in Bull. Herb. Boiss., ser. 2, 7 : 589 (1907) ; Léveillé, Iconogr. Epilobium, t. 49 (1910). Type : Tanganyika, Kilimanjaro, *Volkens* 1847 (G, holo.!, K, iso.!)
E. dolichopodum Samuelsson in N.B.G.B. 9 : 328 (1925). Types : Kenya, Aberdare Mts., Sattimma, *Fries* 2431 (UPS, syn.!) *Fries* 2431a (UPS, lecto.!)

var. **kiwuënse** (*Loes.*) *Brenan* in K.B. 1953 : 163 (1953). Types : Ruanda-Urundi, *Mildbraed* 722, 1555, 1646 (B, syn. †)

Ovaries and capsules, often also young stems, midribs and margins of leaves, pubescent with more or less spreading hairs ; indumentum about 0·25–0·5 mm. high. Fig. 1/3A, (p. 3).

Uganda. Kigezi District : Behungi, 23 Dec. 1933, *Thomas* 1197!
Kenya. Trans-Nzoia District : Mau Range N. of Timboroa, 2 June 1948, *Hedberg* 1080! ; Kiambu District : Limuru, Feb. 1915, *Dümmer* 1644!
Tanganyika. Kilimanjaro, Kibo, 21 Feb. 1933, *C. G. Rogers* 580! ; Pare District : W. of Tona, 13 Feb. 1915, *Peter* K. 659 (O.III.50)! Pare District : Between Tona and Muhasi swamp, 13 July 1915, *Peter* K. 661 (O.III.142)!
Distr. U2 ; K3, 4 ; T2, 3 ; Belgian Congo and Ruanda-Urundi

Syn. *E. kiwuënse* Loes. in Mildbr., Z.A.E. 588 (1913) ; Robyns, F.P.N.A. 1 : 682, t. 72 (1948)

Hab. (of species as whole). By streams and rivers and in swampy places in moor grasslands and upland moor ; alt. of var. *stereophyllum* 2700–3660 m., of var. *kiwuënse* 1750–2740 m.

The habit of *E. stereophyllum* varies much, probably due to altitude and exposure. Thus within var. *stereophyllum, Volkens* 1847! (see above) with comparatively long internodes and wide leaves contrasts strongly with *Adamson* AJ. 491! (see above) with short internodes and narrow leaves, but collected at a much higher altitude.
  In our area there is a tendency, though an inconstant one, for var. *kiwuënse* to have more ovate leaves than var. *stereophyllum*, though in Abyssinia var. *stereophyllum* occurs with ovate leaves. Here too perplexing plants occur with leaves more cuneate at the base (e.g. *Schimper* 544!) possibly due to habitat, but similar plants have not been collected in our area so far.
  Attention is drawn to the interesting difference in range of altitude between var. *stereophyllum* and var. *kiwuënse*.

4. **E. salignum** *Hausskn.* in Oesterr. Bot. Zeitschr. 29 : 90 (1879) ; Hausskn., Monogr. Epilobium 236 (1884) ; Perrier de la Bâthie in Humbert, Not. Syst. 13 : 138 (1947). Type : Madagascar, Antananarivo & Be'zon-zong, *Bojer* (P, lecto., K, photo.!)

Herbaceous or thinly woody, up to 1·5 m. high, often branched above, appressed grey-puberulous on stems, margins and veins of leaves, ovaries, outside of sepals and fruits ; leaves sometimes glabrous ; no capitate glandular hairs. Stems without imbricate scale-leaves at base. Leaves lanceolate to oblong-linear, 2–8 cm. long, 0·3–1·4 cm. wide, obtuse or subacute at apex, cuneate to narrowly rounded at base ; petiole usually 1·5–4·5 mm. long. Flowers nodding, white or cream, often pink-tinged later. Sepals 4·5–7·5 mm. long. Petals 6–11·5 mm. long, 3–4·5 (sometimes to 7) mm. wide, lamina elliptic or narrowly obovate-elliptic. Style 3·5–7 mm. long. Stigma variable in shape, normally oblong to ellipsoid more rarely obovoid, entire or nearly so, 2–3·5 mm. long, 0·7–1·5 mm. wide. Capsules 3–8 cm. long ; pedicels 1·3–4·5 cm. long. Fig. 1/4, (p. 3).

UGANDA. Toro District : E. Ruwenzori, Bwamba Pass, July 1940, *Eggeling* 4031 ! ; Kigezi District : Kachwekano Farm, Jan. 1950, *Purseglove* 3188 !
KENYA. Elgon, Cherangani Hills, Dec. 1933, *Mrs. Powles* 35 ! ; Uasin Gishu District : Kipkarren, Oct. 1931, *Mrs. Brodhurst-Hill* 505 !
TANGANYIKA. Ufipa District : Ufipa Plateau, Lua, 16 Apr. 1934, *Michelmore* 1029 ! ; Uluguru Mts., Lukwangule, 2 Jan. 1934, *Michelmore* 880 !
DISTR. U2 ; K3, 4 ; T4, 6, 7 ; eastern side of Africa from Abyssinia southwards to South Africa, westwards to the French and British Cameroons and Angola ; also in Madagascar
HAB. Marshy and swampy places by streams ; 1070–2830 m.

SYN. E. *neriophyllum* Hausskn. in Abh. Naturw. Verein. Bremen 7 : 19 (1880), *in obs.* ; Hausskn., Monogr. Epilobium 236 (1884) ; Engl. in V.E. 3 (2) : 773 (1921). Type : South Africa, Boschberg, *MacOwan* 1487 (K, lecto. !)
E. *benguellense* [Welw. ex] Hiern, Cat. Afr. Pl. Welw. 1 : 378 (1898). Types : Angola, Huila, *Welwitsch* 4458, 4459 (BM, syn. !, K, isosyn. !)
E. *neriophyllum* Hausskn. subsp. *benguellense* (Welw. ex Hiern) Engl. in V.E. 3 (2) : 773 (1921)
E. *neriophyllum* Hausskn. subsp. *benguellense* (Welw. ex Hiern) Engl. var. *welwitschii* Engl. in V.E. 3 (2) : 773 (1921). No type cited
E. *neriophyllum* Hausskn. subsp. *cylindrostigma* Engl. in V.E. 3 (2) : 773 (1921). No type indicated, but recorded from Tanganyika Territory, Ruanda-Urundi and the French Cameroons
E. *neriophyllum* subsp. *ellenbeckii* Engl. in V.E. 3 (2) : 773 (1921). No specimen given, but the locality : Abyssinia, Arusi-Galla Land, Jidah, 2600 m.

Habit, amount of branching, width of leaves, shape and size of stigma all vary a good deal in this species, but I can find no evidence to justify the elaborate subdivisions of Engler in V.E. 3 (2) : 773 (1921). The indumentum, leaf-shape and characteristic colour of the rather small flowers together usually make E. *salignum* an easily recognized species. Some Madagascar gatherings look unusual in their rather broad foliage and slightly larger flowers, possibly induced by shade, but agree closely in essentials with continental African specimens.

2. JUSSIAEA

L., Gen. Pl., ed. 5, 183 (1754)

Ludwigia L., Gen. Pl., ed. 5, 55 (1754)

Mostly annual or perennial herbs, rarely becoming shrubby, or very rarely even a small tree. Leaves alternate, entire in the African species. Stipules very small or none. Flowers regular. Sepals mostly 4–5, rarely 3–7, persistent after flowering. Receptacle not or hardly prolonged beyond the ovary. Petals as many as sepals, in our species always yellow (in others sometimes white). Stamens as many as petals (*Ludwigia* L.) or twice as many (*Jussiaea* L., *sens. strict.*). Stigma capitate, subentire or lobulate. Ovules uniseriate or pluriseriate in each cell of the 3–7-celled ovary. Fruit a capsule, normally elongate, sometimes short, septicidally dehiscent with the valve-midribs later splitting, or with the pericarp irregularly breaking, sometimes very slowly so or apparently not. Seeds without hairs, free or surrounded by endocarp ; raphe sometimes much enlarged.

Ludwigia L. is here amalgamated with *Jussiaea*. To separate it on account of the stamens being in one whorl instead of two seems to me to sever closely related species and to cut across their natural affinities, commonly based on the fruit and seeds. It appears that the loss of a whorl of stamens has happened more than once during the evolutionary history of *Jussiaea*, in fact that *Ludwigia* represents a polyphyletic trend.

In the following account I have grouped the species according to the fruit and seed characters :—

A. Seeds pluriseriate, free, narrow-raphed (Sect. *Myrtocarpus* Munz in Darwiniana 4 : 184 (1942)). Species nos. 1–4.

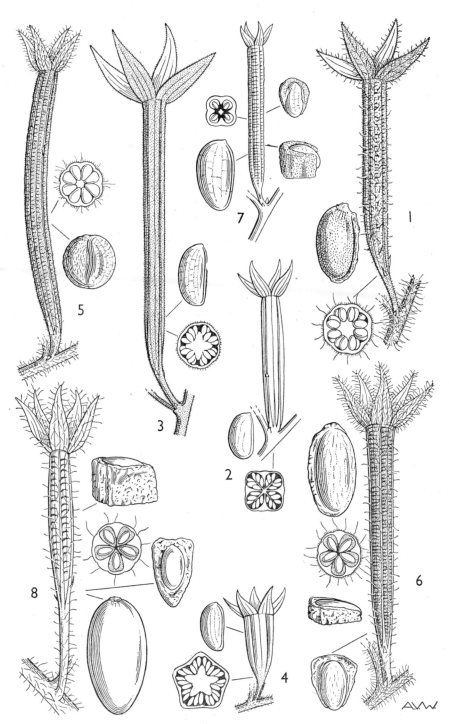

FIG. 2. *JUSSIAEA*—Capsules (external views, × 2, and diagrammatic cross-sections, × 4) and seeds, × 24 ; 6, 7, 8 each show also two views (× 12) of the piece of endocarp surrounding the seed. 1, *J. stenorraphe* var. *stenorraphe*; 2, *J. erecta* ; 3, *J. jussiaeoides*; 4, *J. perennis*; 5, *J. suffruticosa* var. *brevisepala*; 6, *J. leptocarpa*; 7, *J. abyssinica*; 8, *J. repens* var. *diffusa*.

B. Seeds pluriseriate, free, broad-raphed (Sect. *Macrocarpon* Micheli in
    Mart., Fl. Bras. 13 (2) : 149, 169 (1875)). Species no. 5.
C. Seeds uniseriate, partially enclosed by but free within pieces of
    endocarp. Species nos. 6–7.
D. Seeds uniseriate, each surrounded by and firmly attached to its piece
    of endocarp (Sect. *Jussiaea*, Sect. *Eujussiaea* Munz in Darwiniana 4 :
    184 (1942)). Species no. 8.

White to pink, inflated, spongy floating roots produced
 from nodes. . . . . 8. *J. repens*
Inflated floating roots absent from nodes :
 Seeds pluriseriate in each cell of the capsule, free, i.e.
  not enclosed in, or surrounded by, endocarp ;
  sepals mostly 4, if 5 then capsule almost always
  1·2 cm. long or less :
  Seeds each with an enlarged raphe almost as big as
   the body of the seed itself ; otherwise very
   variable . . . . 5. *J. suffruticosa*
  Seeds each with a raphe much narrower than the
   body of the seed :
   Stamens 4 (–5) :
    Sepals 6–13 mm. long ; petals 10–15 mm. long,
     10–14 mm. wide ; capsules 2–4·3 cm. long ;
     seeds 0·5–0·6 mm. long, 0·3–0·4 mm. wide ;
     plant always puberulous or shortly pubes-
     cent, especially on young parts . . 3. *J. jussiaeoïdes*
    Sepals 2–3·5 mm. long ; petals 1–3 mm. long,
     0·75–2 mm. wide ; capsules 0·3–1·2 (–1·5)
     cm. long ; seeds 0·3–0·4 mm. long, 0·2 mm.
     wide ; plant minutely puberulous to
     glabrous . . . . . . 4. *J. perennis*
   Stamens 8 (–10) :
    Plant glabrous (at least in African specimens) ;
     petals 4–5 mm. long ; seeds 0·3–0·5 mm.
     long, 0·2–0·3 mm. wide ; petiole 3–15 mm.
     long ; capsule not exceeding 1·9 cm. in
     length . . . . . . 2. *J. erecta*
    Plant always pubescent or hairy ; petals 6–16
     mm. long ; seeds 0·75–0·8 mm. long, 0·4
     mm. wide ; petiole up to 4 mm. long ;
     capsule often 2–4 cm. long . . . 1. *J. stenorraphe*
 Seeds uniseriate in each cell of the capsule, each
  enclosed or surrounded by a piece of endocarp ;
  sepals 4–5 :
  Sepals 1·75–3 mm. long ; petals 1·5–3·5 mm. long ;
   stamens 4 (–5) ; capsules 1–2 mm. wide ; stems
   and ovaries glabrous . . . . . 7. *J. abyssinica*
  Sepals 5–14 mm. long ; petals 5–24 mm. long ;
   stamens 8–10 (–14) ; capsules 2·5–4 mm. wide :
   Leaves with lateral nerves 6–12 each side of mid-
    rib ; capsules marked on outside with little
    bumps 1·5–2 mm. apart, corresponding to the
    seeds inside ; seeds pendulous, each enclosed
    by and firmly attached to its piece of
    endocarp, the actual seed 1–1·5 mm. long ;
    plant creeping or floating . . . 8. *J. repens*

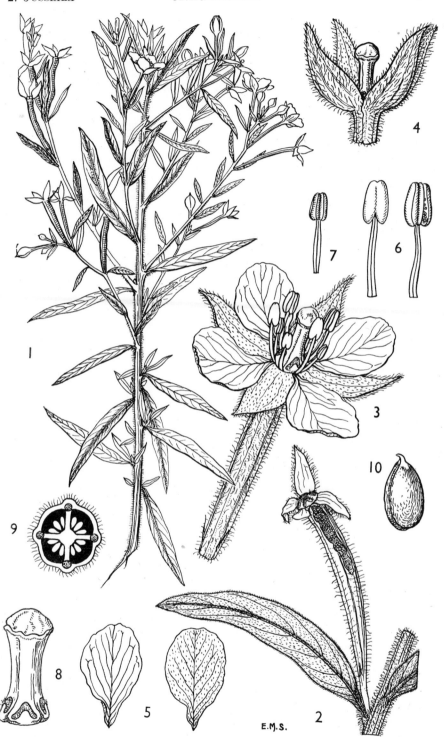

E.M.S.

FIG. 3. *JUSSIAEA STENORRAPHE* var. *STENORRAPHE*—1, part of plant, × 2/3 ; 2, partly dehisced capsule with its subtending leaf, × 2 ; 3, flower, × 3 ; 4, flower, after removal of petals and stamens, × 3 ; 5, petal, inner and outer faces, × 3 ; 6, antepetalous stamen, inner and outer faces, × 6 ; 7, antesepalous stamen, inner face, × 6 ; 8, style and stigma, × 6 ; 9, transverse section of ovary, diagrammatic, × 20 ; 10, seed, × 40.

Leaves with lateral nerves 11–20 each side of midrib ; capsules marked on outside with little bumps 0·5 mm. apart, corresponding to the seeds inside ; seeds horizontal, each surrounded by but free within a horseshoe-shaped piece of endocarp, the actual seed 0·8–1 mm. long . . . . . 6. *J. leptocarpa*

1. **J. stenorraphe** *Brenan* in K.B. 1953 : 164 (1953). Type : Nigeria, 28 km. N. of Kaduna, *Keay FHI.* 28113 (K, holo. !)

Very polymorphic. Shrubby or less often herbaceous, erect, much branched, 1–3 m. high, more or less densely spreading-hairy all over, rarely sparingly so or merely pubescent. Leaves from almost linear to elliptic, 2–10 cm. long, 0·2–3·8 cm. wide, acute or subacute at apex ; petiole 0–4 mm. long, rarely more. Sepals 4, (4–) 6–13 mm. long, 1·5–5 mm. wide. Petals yellow, 6–16 mm. long, 4–16 mm. wide. Stamens twice as many as sepals. Apex of ovary, supporting the 2–6 mm. long style, flat or somewhat raised (to 2 mm.). Capsules thin-walled, puberulous to hairy, 1–4 cm. long, 1·5–4 mm. wide ; pedicels 1–10 (–20) mm. long. Seeds pluriseriate in each cell, free, ellipsoid-oblong, narrow-raphed, 0·75–0·8 mm. long, 0·4 mm. wide, brown.

Key to Intraspecific Variants

Sepals 4–9 (–10) mm. long :
  Plant more or less plentifully spreading-hairy ; capsules about 2–3 cm. long . . . . . var. **stenorraphe**
  Plant puberulous or shortly pubescent ; capsules about 1–1·5 cm. long, poorly formed, most seeds aborting . . . . . var. **reducta**
Sepals (9–) 10–14 mm. long :
  Leaves, at least the cauline ones, with long hairs on both surfaces, but without an " understorey " of short ones ; petals 11–14 mm. long . . var. **macrosepala**
  Leaves, at least the cauline ones, with long hairs on both surfaces and in addition an " understorey " of more or less numerous, much shorter, appressed hairs ; petals about 16 mm. long . . . var. **speciosa**

var. **stenorraphe** ; Brenan in K.B. 1953 : 164 (1953).

Plant more or less plentifully spreading-hairy. Leaves mostly linear to linear-lanceolate, including those subtending flowers. Sepals 6–9 (–10) mm. long. Petals 9·5–13 mm. long, 6–12 mm. wide, obovate. Apex of ovary, supporting the 2–4 (–6) mm. long style, flat or up to 1·5 mm. high. Capsules about 2–3 cm. long ; pedicels 3–10 mm. long. Figs. 2/1, (p. 7) & 3, (p. 9).

Uganda. West Nile District : Madi, Dec. 1862, *Speke & Grant* ! ; Teso District : Serere, Aug. 1932, *Chandler* 841 ! ; Masaka District : Buddu, Kyebe, Aug. 1945, *Purseglove* 1779 !
Tanganyika. Mwanza District : NW. Uzinza, Bugando Chiefdom, 18 June 1937, *Burtt* 6536 ! ; ? Kigoma District : Luiche R. between Kigoma and Machaso, 18 Feb. 1926, *Peter* 37018 !
Distr. U1, 3, 4 ; T1, 4 ; French Guinea, ? Gold Coast, British Togoland, Nigeria, Anglo-Egyptian Sudan, Northern Rhodesia, the Belgian Congo (Katanga) and Angola

The most significant characters of this, the most widespread variety, are the numerous spreading hairs and short sepals. It appears to occur nowhere east of Lake Victoria.

var. **reducta** *Brenan* in K.B. 1953 : 166 (1953). Type : Uganda, Masaka District, *Hansford in Snowden Herb.* 2356 (K, holo. !)

Plant puberulous or shortly pubescent, and with a few rather short spreading hairs less than 1 mm. long on young parts. Leaves mostly small, those subtending flowers lanceolate, to about 3 cm. long and 0·8 cm. wide. Sepals 4–8 mm. long, 1·5–2·5 mm. wide. Petals 6–10 mm. long, 4–8 mm. wide. Apex of ovary, supporting the 2–3 mm. long style, about 0·7–1·5 mm. high. Capsules characteristically poorly formed, 1–1·5 cm. long, 1·5–2 mm. wide, most seeds aborting ; pedicels 1–3 mm. long.

UGANDA. Masaka District : Luwera-Masaka road, Oct. 1931, *Hansford in Snowden Herb.* 2356! ; Mengo District : Namanyonyi, Jan.-Feb. 1916, *Dümmer* 2738!
TANGANYIKA. Lake Province : without precise locality, *Conrads* 615!
DISTR. **U**4 ; **T**1 ; not found elsewhere

The few and inconspicuous spreading hairs combined with short sepals are enough to indicate var. *reducta*, which is evidently more closely related to var. *stenorraphe* than the two following varieties. The poorly formed fruits are present on all the cited gatherings, of which that of Conrads is very copious. At present it is probably better to treat var. *reducta* as merely a local variant endemic to the region of Lake Victoria, but botanists are urged to observe and collect more of it in the field so that its status may be put beyond doubt. I feel that the possibility of its being a hybrid should be borne in mind. Besides the cited gatherings, two others from Mengo District—*Maitland* 537 ! and *Hancock & Chandler* 113 !—are probably var. *reducta*, but collected too young.

Ward 4g ! Tanganyika, **T**7, Iringa District, Idodi, Sept.-Oct. 1936, is very reminiscent of var. *reducta*, but has larger foliage. More material from that locality would be most welcome.

var. **macrosepala** *Brenan* in K.B. 1953: 166 (1953). Type : Uganda, West Nile District, Terego, *Hazel* 665 (K, holo. !)
Plant, especially on young parts, plentifully clothed with spreading hairs 1·5–2 mm. long. Leaves, at least the cauline ones, with long hairs on both surfaces, but without any " understorey " of short ones ; leaves subtending the flowers mostly lanceolate. Sepals mostly 10–14 mm. long and up to 5 mm. wide. Petals almost round, 11–14 mm. long, 10–15 mm. wide. Apex of ovary, supporting the 3·5–5 mm. long style, almost flat to about 1·5 mm. high. Capsules 2–4 cm. long ; pedicels 5–20 mm. long.

UGANDA. West Nile District : Terego, Aug. 1938, *Hazel* 665!
KENYA. Kiambu District : Theta papyrus swamp, *Battiscombe* 1122! ; North Kavirondo District : from Nandi to Mumias, 1898, *Whyte*!
TANGANYIKA. Pare District : Butu, Oct. 1927, *Haarer* 836! ; Pare District : Gonja Swamp, Oct. 1935, *R. M. Davies* 1108! ; Rungwe District : Msombas, 1 Mar. 1933, *R. M. Davies* 903!
DISTR. **U**1 ; **K**4, 5 ; **T**3, 4, 7 ; Anglo-Egyptian Sudan (probable, but not certain) and Nyasaland

Generally var. *macrosepala* is a distinct and rather uniform plant, readily recognizable by its indumentum and long sepals. In addition there is a characteristic tendency to broader leaves than in var. *stenorraphe* ; the petals are commonly larger and the style usually longer.
There are certain specimens lacking fruits (Zanzibar : Zingwe-Zingwe, 21 Jan. 1929, *Greenway* 1098 (K! EA!) ; **T**3, Lushoto District : Lungusa near Amani, 20 Sept. 1908, *Braun* 2089 (EA!) ; Makuyuni, June 1935, *Koritschoner* 590 (EA!)), which suggest the presence of a further variant of *J. stenorraphe* similar to var. *macrosepala* in facies and sepals, but with a short indumentum resembling that of var. *reducta*. Further material of this is required.

var. **speciosa** *Brenan* in K.B. 1953: 167 (1953). Type : Portuguese East Africa, Lugela District, Namagoa Estate, *Faulkner, Kew No.* 401 (K, holo. !)
Indumentum similar to that of var. *macrosepala* but the leaves, at least the cauline ones, villous with rather long hairs, especially on the midrib, but also with an " understorey " of more or less numerous much shorter appressed hairs. Leaves subtending the flowers elliptic or elliptic-lanceolate. Sepals mostly 9–13 mm. long and up to 4·5 mm. wide. Petals almost round, about 16 mm. long and 15–16 mm. wide. Apex of ovary, supporting the 5–6 mm. long style, 1·25–2 mm. high. Capsules about 2·5 cm. long ; pedicels about 9–12 mm. long.

TANGANYIKA. Rufiji District : Rufiji, 3 Jan. 1931, *Musk* 108!
DISTR. **T**6 ; Portuguese East Africa

The var. *speciosa*, which in full bloom must be uncommonly handsome, is obviously related more to var. *macrosepala* than the other varieties, but differs in the characteristically mixed indumentum on the leaves which tend to be even broader, and also in the yet larger petals.

Its apparent restriction to comparatively low altitudes near the coast is notable, but it has been only very rarely collected.

DISTR. (of species as whole). Widespread in tropical Africa (see above under separate varieties), but unknown elsewhere

HAB. (of species as whole). Swamps by rivers and lakes, flooded grassland and sandy river-beds ; the first three varieties from 600–1370 m. ; the var. *speciosa* only collected at 10 m.

J. stenorraphe has been in the past much confused in herbaria with *J. pilosa* H.B.K. (= *J. leptocarpa* Nutt.) and, rather more pardonably, with *J. villosa* Lam. (= *J. suffruticosa* L.). In many ways it is close to *J. suffruticosa*, but differs from all forms of that in the narrow-raphed and usually longer seeds. In addition the more robust and not infrequently shrubby habit of *J. stenorraphe*, the commonly rather large sepals and petals, and the often longer style will separate it from one or another of the forms of *J. suffruticosa*.

The sepals of *J. stenorraphe* often turn bright red inside after the petals have fallen.

2. J. erecta L., Sp. Pl. 388 (1753) ; Munz in Darwiniana 4 : 195 (1942) ; Munz in Hoehne, Fl. Brasilica 41 (1) : 17 (1947). Type : [? America] in *Herb. Linnaeus* (LINN, lecto. !)

Annual herb to 3 m. high, in Africa absolutely glabrous, even under a × 20 lens ; but a variant with ovaries, sepals, midribs and lateral nerves of leaves minutely puberulous occurs in America. Leaves lanceolate, sometimes ovate or elliptic, mostly 2–13 cm. long and 0·5–4·3 cm. wide, acute or subacuminate, rarely obtuse at apex ; petiole 3–15 mm. long. Sepals almost always 4 (5 seen only in one flower), 3–6 mm. long, 1–1·5 mm. wide. Petals yellow, elliptic or obovate, 4–5 mm. long, 2–2·5 mm. wide. Stamens twice as many as sepals. Apex of ovary, supporting the 0·5–1 mm. long style, flat or only slightly raised. Capsules thin-walled, 1–1·9 cm. long, 2·5–5 mm. wide ; pedicels 0–8 mm. long. Seeds pluriseriate in each cell, free, oblong-ellipsoid, narrow-raphed, 0·3–0·4 (–0·5) mm. long, 0·2–0·3 mm. wide, pale brown. Fig. 2/2, (p. 7).

TANGANYIKA. Tanga District : Amboni, June 1893, *Holst* 2811 ! ; Morogoro District : Ngomero, *Speke & Grant* ! ; ? Kilwa District : Singoni, 31 May 1906, *Braun* 1332 !
ZANZIBAR. Pemba, Chake Chake, 20 Oct. 1929, *Vaughan* 836 !
DISTR. **T**3, 4, 6, 8 ; **P** ; widespread in tropical Africa from Senegal, the French Sudan and the Anglo-Egyptian Sudan southwards to Angola, Southern Rhodesia and Portuguese East Africa ; also on the Comoro Islands, Madagascar and the Seychelles ; in America from Florida southwards to Brazil, Paraguay and Bolivia ; apparently quite absent from Asia although commonly recorded, due to misapplying the name *J. erecta* L. to *J. suffruticosa* L.
HAB. Imperfectly known, probably various wet habitats ; Speke and Grant's specimen collected on " river bank " ; near sea-level to 1100 m.

SYN. *J. acuminata* Sw., Prod. Fl. Ind. Occ. 2 : 745 (1800). Type : Jamaica, *Swartz* (S,? holo. Type material seen by Munz)

3. J. jussiaeoïdes (*Desr.*) Brenan in K.B. 1953 : 163 (1953). Type : Mauritius (Ile de France) *J. Martin in Herb. Lamarck* (P, holo., K, photo. !)

Herb, sometimes slightly woody, up to 3 m. high, minutely puberulous or sometimes shortly pubescent, especially on young parts. Leaves lanceolate to almost linear, mostly 2·5–10 (–11) cm. long, 0·2–1·5 (–2·5) cm. wide, mostly acute at apex ; petiole usually distinct, 2–20 mm. long. Sepals 4, 6–13 mm. long, 2–3·5 mm. wide. Petals yellow, broadly obovate, 10–15 mm. long, 10–14 mm. wide. Stamens as many as sepals. Apex of ovary, supporting the 3·5–5 mm. long style, raised and conical, about 1·5–2·5 mm. high. Capsules thin-walled, minutely puberulous, 2–4·3 cm. long, 2–3 mm. wide ;

pedicels 2–8 mm. long. Seeds pluriseriate in each cell, free, more or less ellipsoid, narrow-raphed, 0·5–0·6 mm. long, 0·3–0·4 mm. wide, pale brown. Fig. 2/3, (p. 7).

KENYA. Kilifi District : Mida, Sept. 1929, *Graham* 2099 ! ; Kilifi District : Rabai, Nov. 1933, *Joanna in C.M.* 5977 !
TANGANYIKA. Pare District : Kihurio, Aug. 1928, *Haarer* 1513 ! ; Lushoto District : Mkomazi, 22 Apr. 1934, *Greenway* 3948 ! Pangani District : ; Bushiri Estate, 29 July 1950, *Faulkner* 633 !
ZANZIBAR. Zanzibar, Mbiji, 8 Feb. 1929, *Greenway* 1383 ! ; Zanzibar, 4 Aug. 1950, *Williams* 58 ! ; Pemba, Chake Chake, 25 Oct. 1929, *Vaughan* 835 !
DISTR. K (? 3, record doubtful), 7 ; T3, 6, 8 ; Z ; P ; Portuguese East Africa, Madagascar, Mauritius and the Seychelles
HAB. Swamps and seasonally flooded places by lakes, streams, etc. ; about sea-level to 760 m.

SYN. *Ludwigia jussiaeoïdes* Desr. in Lam., Encycl. Méth., 3 : 614 (1791) ; Oliv. in F.T.A. 2 : 490 (1871) ; Perrier de la Bâthie in Humbert, Not. Syst. 13 : 141 (1947)

This species seems restricted, so far as our area is concerned, to places on or not far from the coast, and is thus unknown from Uganda. It rather resembles *J. suffruticosa*, differing in the stamens four not eight, the narrow- not broad-raphed seeds, and in the more conical apex to the ovary. From *J. stenorraphe* it differs again in the stamen-number, the shorter seeds and the usually longer petiole.

4. J. perennis (*L.*) *Brenan* in K.B. 1953 : 163 (1953). Type : Ceylon, *P. Hermann* (BM, lecto.!)

Annual herb up to 50 cm. high, minutely puberulous on young parts (× 10 lens necessary) becoming almost glabrous, sometimes altogether glabrous. Leaves lanceolate to linear, mostly 1·5–9 cm. long, 0·15–1·5 cm. wide, subacute or acute at apex ; petiole (2–) 5–15 mm. long. Sepals 4–5, 2–3·5 mm. long, 0·75–1·75 mm. wide. Petals small, yellow, about 1–3 mm. long and 0·75–2 mm. wide, elliptic. Stamens as many as sepals. Apex of ovary, supporting the 0·75–1·5 mm. long style, flat or slightly (to 0·3 mm.) raised. Capsules thin-walled, glabrous or very minutely puberulous, 3–12 (–15 in Asia) mm. long, 2–4 mm. wide at apex ; pedicels 1–6 mm. long. Seeds pluriseriate in each cell, free, more or less ellipsoid, narrow-raphed, 0·3–0·4 mm. long, 0·2 mm. wide, brown to purplish or pinkish. Fig. 2/4, (p. 7).

UGANDA. Bunyoro District, Nov. 1862, *Speke & Grant* !
KENYA. District uncertain : N. of Mombasa, to Lamu and Witu, 1902, *Whyte* !
TANGANYIKA. Shinyanga District : Shinyanga, Mar.–Apr. 1933, *Bax* 52 ! ; Uzaramo District : Dar es Salaam, Mburahati, 12 Sept. 1926, *Peter* 45177 ! ; Ufipa District : Lake Rukwa flood-plain, Milepa, 21 Apr. 1936, *Lea* LR35 !
DISTR. U4 ; K7 ; T1, 4, 6 ; Senegambia, Chad, Anglo-Egyptian Sudan, Portuguese East Africa and Madagascar ; widely distributed through tropical Asia to China (Hainan), the Philippines and New Caledonia ; Australian specimens may belong, but have sessile capsules.
HAB. Swamps, flood-plains of lakes, and probably in other wet habitats ; altitude-range uncertain as records are so few, but probably at least from near sea-level to 820 m.

SYN. *L. perennis* L., Sp. Pl. 119 (1753), excl. verba falsa " foliis oppositis "
L. parviflora Roxb., Fl. Ind. 1 : 440 (1820) ; Oliv. in F.T.A. 2 : 490 (1871). Type from Bengal ; whereabouts of specimen uncertain, but I have seen Roxburgh's own painting of his species (No. 1340) at Kew
Isnardia multiflora Guill. & Perr. in Guill., Perr. & A. Rich., Fl. Senegamb. Tent. 1 : 295 (1832–3). Type : Senegambia, Richard Toll, *Leprieur* (G, holo.!)
L. multiflora (Guill. & Perr.) Walp., Repert. 2 : 75 (1843)

The African specimens are all puberulous except for Leprieur's from Senegambia and one from the Sudan ; they are also all comparatively short-capsuled. In Asia the plant is more polymorphic, and specimens from there with capsules as long as 15 mm. are probably referable to *L. perennis* in a wide sense.

5. **J. suffruticosa** *L.*, Sp. Pl. 388 (1753) ; Munz in Darwiniana 4 : 235 (1952) ; Munz in Hoehne, Fl. Brasilica 41 (1) : 35 (1947). Type : " Jussiaea erecta villosa, floribus tetrapetalis octandris pedunculatis . . . India " (L., l.c.)

Excessively polymorphic and variable. Herb or shrub up to 3·6 m. high, puberulous and subglabrous to densely spreading-hairy. Leaves linear to lanceolate or ovate- or obovate-elliptic, variable in size, normally acute to subacuminate at apex ; petiole of upper and median leaves normally short, 0–4 mm. long, sometimes longer in lower cauline leaves. Sepals 4, 3–19 mm. long, 1–10 mm. wide. Petals yellow, 3·5–28 mm. long, 2–28 mm. wide. Stamens twice as many as sepals. Apex of ovary, supporting the 1·5–3·5 mm. long style, flat or slightly raised (to about 1 mm. high). Capsules thin-walled, puberulous to hairy, 2–5·5 cm. long, 2–6 mm. wide ; pedicels 0–20 (or more) mm. long, in Africa normally less than 7 mm. long. Seeds pluriseriate in each cell, free, each with an enlarged raphe almost as big as the body of the seed, together forming a more or less round brown body 0·5–0·75 mm. in diameter, with a marked furrow running up its middle.

KEY TO INTRASPECIFIC VARIANTS

Sepals 3–6 mm. long :
    Leaves mostly lanceolate to linear-lanceolate, those
        subtending flowers 2–10 mm. wide, gradually
        tapering towards apex, not parallel-sided
                        subsp. **suffruticosa** var. **brevisepala**
    Leaves linear, those subtending flowers 1–2 (–3) mm.
        wide, more or less parallel-sided :
        Leaves and stems puberulous, appearing sub-
            glabrous, sometimes with a few spreading
            hairs on young parts only
                        subsp. **suffruticosa** var. **linearis**
        Leaves and stems more or less plentifully clothed
            with long spreading hairs
                        subsp. **suffruticosa** var. **piloso-linearis**
Sepals 6–13 mm. long :
    Leaves linear ; plant puberulous or with spreading
        hairs on young parts only
                        subsp. **suffruticosa** var. **linearifolia**
    Leaves obovate-elliptic to obovate-lanceolate ; stems
        and leaves more or less densely clothed with
        rather long spreading hairs
                        subsp. **octonervia** var. **sessiliflora**

subsp. **suffruticosa** ; see note on p. 16.

Habit normally herbaceous, sometimes slightly woody, to about 1·2 m. high. Leaves linear to lanceolate. Sepals lanceolate to narrowly ovate-elliptic, up to about 4·5 mm. wide. Petals up to about 10 mm. long and 8 mm. wide. Anthers normally about 0·5–1·6 mm. long. Capsule tending to be more cylindrical and less tapering than in subsp. *octonervia*, and to be shorter than its subtending leaf.

subsp. **suffruticosa** var. **brevisepala** *Brenan* in K.B. 1953 : 168 (1953). Type : ? French Cameroons, Cameroon River, *Mann* 2227 (K, holo. !)

More or less spreading-hairy, sometimes densely so, on stems, leaves and young shoots, or sometimes the spreading hairs on young parts only. Leaves mostly lanceolate to linear-lanceolate, those subtending flowers 2–10 mm. wide, gradually tapering towards apex, not parallel-sided. Sepals 3·5–6 mm. long, 1·5–3·25 mm. wide. Petals 5–8 mm. long, 3–4·5 mm. wide. Fig. 2/5, (p. 7).

UGANDA. West Nile District : Maracha, Dec. 1939, *Hazel* 400!
KENYA. Teita District : Voi, 9 May 1931, *Napier* 999!
TANGANYIKA. Shinyanga District : Shinyanga, *Koritschoner* 3005!; Lushoto District : Mashewa, *Holst* 8787!
ZANZIBAR. Zanzibar, July–Nov. 1873, *Hildebrandt* 972!; Zanzibar, *Mrs Taylor* 80! Pemba, Chake Chake, 20 Oct. 1929, *Vaughan* 837!
DISTR. U1, K7 ; T1, 3, 6 ; Z ; P ; widespread in tropical Africa from Sierra Leone, Uganda and the Anglo-Egyptian Sudan to Southern Rhodesia and Angola

It seems, from the description, quite likely that *J. didymosperma* Perr. in Humbert, Not. Syst. 13 : 148 (1947), from Madagascar is nothing else than *J. suffruticosa* var. *brevisepala*, but I have not seen specimens from there ; except for this the var. *brevisepala* does not seem to occur outside continental Africa. It varies a good deal both in habit and indumentum, and intergrades with the two following varieties.
The most significant characters are the short narrow sepals and the leaf-shape.

subsp. **suffruticosa** var. **linearis** (*Willd.*) [*Oliv. ex*] *O. Ktze.*, Rev. Gen. Pl. 1 : 251 (1891). Type : Guinea, *Isert in Herb. Willdenow* (B, holo. !)

Puberulous, appearing subglabrous, sometimes with a few spreading hairs on young parts only. Leaves linear, those subtending flowers 1–2 (–3) mm. wide, more or less parallel-sided. Sepals 3–5 mm. long, 1–2 mm. wide. Petals 3–4 mm. long, 2–2·5 mm. wide.

UGANDA. Busoga District : Gadumire, July 1926, *Maitland* 1089!; Teso District : Serere, Dec. 1931, *Chandler* 183!
TANGANYIKA. Mwanza District : Mwanza, Aug. 1932, *Rounce* 210!; Mwanza District, 23 Feb. 1933, *Wallace* 649!; Tabora District : Tabora, Mar. 1939, *Lindeman* 655!
DISTR. U3 ; T1, 4 ; Senegal and the Anglo-Egyptian Sudan southwards to Northern Rhodesia and Portuguese East Africa ; also in Madagascar (*Forbes*!, BM) ; not found outside Africa

SYN. *J. linearis* Willd., Sp. Pl. 2 : 575 (1800)
    J. nodulosa Peter in Abh. Gesellsch. Wiss. Göttingen, Math. Phys. Kl., n.f., 13 (2) : 88 (1928). Types : Tanganyika, Western Province, Peter 35476, 35608, 35758, 36111, 36409, 36446, 46167 (B, syn. !)
    J. linearis Peter, l.c. 27 (1928), *nomen nudum* (= *J. nodulosa* Peter)

Typical var. *linearis* is a well-marked and rather uniform plant with characteristic indumentum, very narrow leaves and small flowers. The smaller petals than in var. *brevisepala* are noteworthy. Intermediates, however, occur.
Its absence from the coastal areas of East Africa, in contrast to var. *brevisepala*, is interesting, and suggests that the variants are more than phenotypically different.

subsp. **suffruticosa** var. **piloso-linearis** *Brenan* in K.B. 1953 : 169 (1953). Type : Togo, *Baumann* 246 (K, holo. !)

Similar in size and shape of leaves and size of sepals and petals to var. *linearis*, but leaves and stems more or less plentifully clothed with long spreading hairs.

KENYA. Northern Frontier Province : Dandu, 20 Mar. 1952, *Gillett* 12603 ! District doubtful : River Athi, 20 miles beyond Thika, 1 May 1939, *Bally in C.M.* 9202 !
TANGANYIKA. Pare District : Kisiwani, 14 July 1915, *Peter K* 662 !
DISTR. K1, 4 ; T3 ; Togo, Nigeria and Belgian Congo ; not found outside Africa

This is the least certain and satisfactory of the East African varieties. Further study and observation is wanted here. The first Kenya locality is at 880 m., the second is at the upper limit of altitude, so far as is known, for *J. suffruticosa* in East Africa—1280 m. ; that in Tanganyika at 650 m.

subsp. **suffruticosa** var. **linearifolia** *Hassler* in F.R. 12 : 277 (1913). Type : Paraguay, Camp San Luis, *Fiebrig* 4128 *in Herb. Hassler* (G, ? holo.)

Puberulous or with a few spreading hairs on young parts only. Leaves linear, those subtending flowers 1–4 (–5) mm. wide. Sepals 6–11 mm. long and 2–4 mm. wide. Petals 6–9 mm. long, 4–4·5 mm. wide.

TANGANYIKA. Mwanza District : Mwanza, *R. L. Davis* 268!; Pangani District : Hale and the Pangani Falls, 9 June 1914, *Peter K* 653 !
DISTR. T1, 3, 6 ; Portuguese East Africa, Nyasaland, Southern Rhodesia and Natal ; in the New World extending from Florida to the Argentine ; an Australian specimen (*Hubbard* 2789 !) seems referable here.

SYN. *J. suffruticosa* L. var. *ligustrifolia* (H.B.K.) Griseb. f. *linearifolia* (Hassler) Munz in Darwiniana 4 : 243 (1942) ; Munz in Hoehne, Fl. Brasilica 41 (1) : 38 (1947)

This is very distinct among the African variants of *J. suffruticosa* by the combination of linear leaves and long sepals. In Africa it has a markedly south-eastern distribution, entirely absent from West Africa, but extending as far south as Natal, where it and var. *sessiliflora*, poles apart in appearance, alone represent the species.

A specimen from Kenya, **K4**, Embu District : Itabwa, 6 Apr. 1932, *Sunman* 2219 !, is close to var. *linearifolia* in facies but has the sepals 5–7 mm. long and 2–3·25 mm. wide. Whether this is an unusual form of var. *linearis* or of var. *linearifolia* must be decided by further collection in that area.

I am here following Munz's interpretation of var. *linearifolia*, except that I prefer to keep it as a distinct variety and not as a form of var. *ligustrifolia* (H.B.K.) Griseb., which does not appear to occur in Africa.

subsp. **octonervia** (*Lam.*) *Hassler* in Bull. Soc. Bot. Genève, ser. 2, 5 : 271 (1914). Type from West Indies, in *Herb. Lamarck* (P, holo.)

Habit commonly shrubby, up to 3·6 m. high, but sometimes herbaceous. Leaves broadly lanceolate to obovate-elliptic or ovate-elliptic. Sepals broadly ovate, 4–10 mm. wide. Petals exceeding 9 mm. long and 8 mm. wide. Anthers normally 2–3 mm. long. Capsule tending to taper from near apex to base, and to exceed its subtending leaf.

SYN. *J. octonervia* Lam., Encycl. Méth. 3 : 332, t. 280, fig. 1 (1789)

subsp. **octonervia** var. **sessiliflora** (*Micheli*) *Hassler* in Bull. Soc. Bot. Genève, ser. 2, 5 : 271 (1914) ; Munz in Darwiniana 4 : 237 (1942) ; Munz in Hoehne, Fl. Brasilica 41 (1) : 36 (1947). Type : Brazil, Rio de Janeiro, *Burchell* 927 (K, lecto. *fide* Munz, but I have been unable to find the specimen)

Stems and leaves, especially when young, more or less densely hairy with spreading hairs about 1·5–2 mm. or more long. Leaves obovate-elliptic to obovate-lanceolate. Sepals 8–13 mm. long, 4·5–7 mm. wide.

ZANZIBAR. Pemba : Kipangani, 18 Feb. 1929, *Greenway* 1474 !
DISTR. **P** ; Southern Rhodesia, Natal, Madagascar, Mauritius, Seychelles, India, New Caledonia, Fiji, New Hebrides ; in the New World from Trinidad and Tobago to Brazil, where it is common

SYN. *J. ovalifolia* Sims in Bot. Mag. 52 : t. 2530 (1825). Type : a cultivated plant grown from seeds received from Madagascar.
J. octonervia Lam. f. *sessiliflora* Micheli in Mart., Fl. Bras. 13 (2) : 171, t. 35 (1875).

DISTR. (of species as whole). Widespread in the tropics and subtropics of the Old and New Worlds.
HAB. (for species as a whole). Damp and swampy places by streams, rivers, lakes, etc. ; 0–1280 m.

Micheli, Note sur les Onagrariées du Brésil (Arch. Sci. Bibl. Universelle) (1874), argued that *J. octonervia* Lam. should be a distinct species, a view followed more recently by Jonker (Fl. Suriname) and Cheesman (Fl. Trinidad and Tobago). I was tempted to accept it too, especially as var. *sessiliflora* is so abundantly distinct from the other African variants of *J. suffruticosa*. The specific characters assigned to *J. octonervia* by recent authors are inconstant, however, and both in America and Asia there are numerous very perplexing intermediates between subsp. *suffruticosa* and subsp. *octonervia*. The var. *sessiliflora* is the plant described as *J. suffruticosa* L. by Perrier de la Bâthie in Humbert, Not. Syst. 13 : 147 (1947).

J. suffruticosa in fact seems to be a species evolutionarily in full cry, so to speak, and would surely be an almost ideal subject for cytogenetical work in the tropics.

In interpreting *J. suffruticosa*, I follow the reasonable view of Fawcett and Munz. Linnaeus' conception of this was clearly a muddle, and there is no specimen in his herbarium. It is thus reasonable to typify *J. suffruticosa* by Linnaeus' descriptive phrase, evidently derived from a plant he had seen, and by the habitat, India. These can hardly refer to any other plant but the present, though there is nothing actually to exclude *J. peruviana* L., which is unlikely to have been in India so early.

6. J. leptocarpa *Nutt.*, Gen. N. Amer. Pl. 1 : 279 (1818) ; Munz in Darwiniana 4 : 254 (1942) ; Munz in Hoehne, Fl. Brasilica 41 (1) : 43 (1947). Type : United States, Mississippi, *Nuttall* (PH, holo.)

FIG. 4. *JUSSIAEA ABYSSINICA*—1, part of plant, natural size ; 2, flower, × 4 ; 3, sepal, × 6 ; 4, petal, × 6 ; 5, flower after removal of sepals and petals, × 6 ; 6, stamen (inner face), × 6 ; 7, stamen (outer face), × 6 ; 8, flower bud, × 4 ; 9, bracteole, × 12 ; 10, calyx seen from above, × 6 ; 11, seed with its piece of endocarp, × 10 ; 12, seed removed from its endocarp, × 10 ; 13, fruit before dehiscence, × 2 ; 14, 15, stages in dehiscence of fruit, × 2 and × 4, respectively.

Herb, sometimes slightly woody, 0·4–2 m. high, with more or less numerous sometimes dense spreading hairs especially on young parts, also with an " understorey " of more or less dense very short puberulous hairs (variants occur in the New World lacking the longer hairs or even completely glabrous). Leaves normally lanceolate to sometimes elliptic, variable in size according to position, mostly 3·5–15 cm. long, 1–4 cm. wide (only 2–3 mm. wide in a linear-leaved Cuban variant), usually acute at apex, lateral nerves 11–20 each side of midrib ; petiole 2–20 mm. long. Sepals usually 5, sometimes 4, 6 or even 7, 5–10 mm. long, 1·5–2·25 mm. wide. Petals yellow, obovate, 5–11 mm. long, 4·5–8 mm. wide. Stamens twice as many as sepals. Apex of ovary, supporting the 3–4·5 mm. long style, flat or only slightly (to 0·75 mm.) raised. Capsules thin-walled, slowly dehiscent, 1·5–4·5 (–5) cm. long, 2·5–4 mm. wide, marked on the outside with little bumps about 0·5 mm. apart, corresponding to the seeds ; pedicels 0·2–2 cm. long. Seeds uniseriate in each cell, horizontal, each surrounded by but free within a horseshoe-shaped piece of powdery brown endocarp about 1–1·5 mm. long and 1 mm. wide, the actual seed flattened oblong-ellipsoid, narrow-raphed, 0·8–1 mm. long, 0·5 mm. wide, pale brown. Fig. 2/6, (p. 7).

UGANDA. Bunyoro District : Mutunda, Apr. 1943, *Purseglove* 1560! ; Mengo District : Entebbe, 1903, *Dawe* 18! ; Mengo District : Kampala, King's Lake, 5 Dec. 1935, *Hancock & Chandler* 114!
TANGANYIKA. Bukoba District : Bukoba, Mar. 1932, *Haarer* 2493! ; Dodoma District : Chaya Lake, 19 Sept. 1931, *Burtt* 1358! ; Tundura District, *Allnutt* 38 !
ZANZIBAR.   Oct. 1872, *Hildebrandt* 970 !
DISTR.   **U**4 ;   **T**1, 4, 5, 8 ;   **Z** ;   widely distributed in tropical Africa from Sierra Leone and the Anglo-Egyptian Sudan southwards to Portuguese East Africa, Angola and South-West Africa ; in America from Georgia and Florida to the Argentine ; absent from Asia
HAB.   Swamps, ditches, rivers, ponds and lakes ; less than 120–1280 m.

SYN.   *J. pilosa* H.B.K., Nov. Gen. & Sp. Pl. 6 : 101, t. 532a & b (1823). Type : Venezuela, R. Apure near San Fernando, *Humboldt & Bonpland* (P, holo.)

Our plants are relatively constant, and all referable to *J. leptocarpa* var. *leptocarpa* (*J. leptocarpa* var. *genuina* Munz in Darwiniana 4 : 255 (1942)).
In the nomenclature of this species I follow Munz, who has evidently seen Nuttall's type-specimen, and also material of *J. pilosa* from Humboldt's herbarium.
Although *J. leptocarpa* is unrecorded for Kenya, it must surely occur on the Kenya shore of Lake Victoria, and should be looked for there.
J. leptocarpa is often confused, at any rate in herbaria, with hairy variants of *J. repens*, but is easily separable by the absence of pneumatophores from the nodes, the more numerous lateral nerves of the leaf, and by the horizontal not vertical seeds showing through the capsule wall as bumps much closer together than in *J. repens*. From the base of the sinus between each sepal of *J. leptocarpa* a pronounced ridge, almost a narrow wing, runs down the apical part of the ovary and young fruit ; in *J. repens* this is almost absent.
I suspect that *J. seminuda* Perr. in Humbert, Not. Syst. 13 : 146 (1947) is *J. leptocarpa* Nutt., and that the plant there described as *J. pilosa* H.B.K. is something quite different.

7. **J. abyssinica** (*A. Rich.*) *Dandy & Brenan* in F. W. Andrews, Fl. Pl. A.-E. Sudan 1 : 145 (1950). Type : Abyssinia, Shire, *Quartin-Dillon & Petit* (P, holo.!)

Herb or weak shrub, straggling or erect, up to 3 m. high, much branched, absolutely glabrous on young parts (under a × 20 lens) except often for extremely minute hairs on margins and midribs of leaves. Stems often red. Leaves lanceolate to elliptic-lanceolate, rarely elliptic, variable in size, mostly 2–13 cm. long and 0·5–2·7 (–3·5) cm. wide, acute or subacute at apex ; petiole 2–20 mm. long. Flowers characteristically appearing as though in axillary clusters. Sepals 4 (? very rarely 5 *), 1·75–3 mm. long, 0·5–1 mm.

* Although Richard described the sepals and stamens of *Ludwigia abyssinica* as 4–5, I cannot see any 5-merous flowers on the type. Richard's statement is my only evidence for this variation.

wide. Petals small, yellow, broadly obovate, 1·5–3·5 mm. long, 1·2–2 mm. wide. Stamens as many as sepals. Apex of ovary, supporting the 0·5–0·75 mm. long style, only slightly raised, about 0·5 mm. high. Capsules thin-walled, glabrous, (10–) 13–18 (–20) mm. long, 1–2 mm. wide ; pedicels 0·5–3 mm. long. Seeds uniseriate in each cell, horizontal, each enclosed in a more or less horseshoe-shaped piece of brown powdery endocarp 0·6–1 mm. long and 0·5–0·7 mm. wide, the actual seed ellipsoid, brown, 0·6–0·75 mm. long and 0·4–0·5 mm. wide. Figs. 2/7, (p. 7) & 4.

UGANDA. West Nile District : Maracha, Dec. 1939, *Hazel* 405 ! ; Mbale District : Bugishu, Budadiri, Jan. 1932, *Chandler* 424 ! ; Mengo District : Busiro, Kyewaga Forest near Entebbe, 21 Sept. 1949, *Dawkins* 386 !
KENYA. Uasin Gishu District : Kipkarren, Mar. 1932, *Brodhurst-Hill* 689 ! ; Nairobi, Thompson's Estate, 24 Sept. 1930, *Napier* 422 !
TANGANYIKA. Lushoto District : Amani, 13 May 1950, *Verdcourt* 196 ! ; Morogoro District, 20 Nov. 1932, *Wallace* 469 ! ; Rungwe District: Bulambia, 25 Oct. 1912, *Stolz* 1629 !
DISTR. U1–4 ; K3–5 ; T1–7 ; Sierra Leone, the Anglo-Egyptian Sudan and Abyssinia southwards to Angola, Natal and Zululand ; also in Madagascar
HAB. Swampy ground by ditches, rivers and lakes ; 580–1980 m.

SYN. *Ludwigia abyssinica* A. Rich., Tent. Fl. Abyss. 1 : 274 (1848)
[*Ludwigia prostrata* sensu Oliv. in F.T.A. 2 : 491 (1871) ; Perrier de la Bâthie in Humbert, Not. Syst. 13 : 141 (1947), *non* Roxb.]

This plant has been regularly confused with the Asiatic *Ludwigia prostrata* Roxb., which is indeed uncommonly similar in general facies, but has the young parts of the plant puberulous, the seeds free and not encased in pieces of endocarp, and 4-angled fruits only about 1 mm. thick.

8. J. repens *L.*, Sp. Pl. 388 (1753) ; Munz in Darwiniana 4 : 270 (1942). Type : India, *Herb. Linnaeus* (LINN, lecto. !)

Herb, creeping and rooting at nodes or more often floating and then normally producing clusters of whitish to pinkish, inflated, spongy, fusiform, floating roots from the nodes (this is the only East African species in the genus to produce them thus) ; plant glabrous to sometimes more or less densely spreading-pubescent. Leaves variable in shape, spathulate to linear-lanceolate, mostly 1–10 cm. long, 0·7–4 cm. wide, rounded and emarginate to acute at apex, lateral nerves 6–12 each side of midrib ; petiole 2–50 mm. long. Sepals 4–5, 5–14 mm. long, 1·5–2·75 mm. wide. Petals yellow (always so in Africa) or white, obovate or sometimes elliptic, 7–24 mm. long, 4–10 (or ? more) mm. wide. Stamens twice as many as sepals. Apex of ovary, supporting the 4–8 mm. long style, flat or nearly so. Capsules firm-walled, not or only very slowly dehiscent, 1·2–4 cm. long, 2·5–4 mm. wide, marked on the outside with little bumps 1·5–2 mm. apart, corresponding to the seeds ; pedicels 0·5–8 cm. long. Seeds uniseriate in each cell, pendulous, each enclosed in and firmly attached to a more or less quadrate piece of endocarp about 1·5 mm. long and wide, the actual seed 1–1·5 mm. long and 0·5–0·7 mm. wide.

var. diffusa (*Forsk.*) *Brenan* in K.B. 1953: 171 (1953). Type : Egypt, Rosetta on the banks of the Nile, *Forskål* (C, holo. !)

Leaves on flowering shoots normally lanceolate to linear-lanceolate, 0·2–1·7 (–2·2) cm. wide, gradually tapering to a normally acute or subacute sometimes obtuse apex ; petiole 2–20 mm. long. Petals yellow, 7–18 mm. long, 4–10 mm. wide. Capsules to 3 cm. long, on pedicels 0·5–2 cm. long. Fig. 2/8, (p. 7).

UGANDA. Acholi District : Gulu, Feb. 1938, *Chorley* 2158 ! ; Masaka District: Bukasa, Sese Islands, 7 June 1932, *Thomas* 148 !
KENYA. Nairobi, Mbagathi road, Ngong R., 2 Jan. 1931, *Napier* 732 ! ; Machakos District : Fourteen Falls on Athi R., 18 Apr. 1948, *Bogdan* 1567 !

Tanganyika. Bukoba District : Bukoba, Mar. 1931, *Haarer* 2497! ; Dodoma District : Manyoni, 22 Aug. 1931, *Burtt* 3396! ; Mafia Island : Kiwawe, 1 Sept. 1931, *Greenway* 5201!

Zanzibar. Pemba, NW. of Jombangome, 18 Dec. 1930, *Greenway* 2755!

Distr.   U1–4 ; K3, 4, 7 ; T1–6, 8 ; P ; Iraq, Syria, Palestine and North Africa ; throughout Africa southwards to the Transvaal and Natal, very doubtful in Madagascar and Mauritius ; certain yellow-flowered Australian specimens are close to var. *diffusa*, but require further study.

Hab.   Swamps, pools, lakes and rivers, usually aquatic and floating ; 10–1830 m.

Syn.   *J. diffusa* Forsk., Fl. Aegypt.-Arab. 210 (1775) ; Oliv. in F.T.A. 2 : 488 (1871)

Distr. (of species as a whole). Widespread in the tropics and warmer regions of the Old and New Worlds.

I follow Munz's treatment of *J. repens* in Darwiniana 4 : 270 *sqq.* (1942) which allows the species a wide range of variation in leaf-shape and flower-colour, and which I feel sure is reasonable and justified.

J. repens var. *repens*, of which I have seen the Linnaean type, is an Asiatic plant with floating roots, normally spathulate leaves rounded or obtuse at the apex, and white flowers, and does not occur in Africa ; it is difficult to judge the constancy of the white flowers from the inadequate field-notes of most collectors in Asia.

J. repens var. *diffusa* is in facies closer to the yellow-flowered *J. repens* var. *peploïdes* (H.B.K.) Griseb., which is widespread in the New World (see Munz in Darwiniana 4 : 272–6 (1942) ; var. *diffusa* differs in the production of floating roots, the leaves more gradually tapering above, and a tendency for the leaves to be somewhat larger.

The very remarkable floating roots or pneumatophores from the nodes are apparently peculiar to *J. repens*, among our species. For an account of the nature and anatomy of these wonderful organs see C. Martins, Mémoire sur les racines aérifères . . . du genre *Jussiaea* . . . (Mém. Acad. Sci. Lettres Montpellier) 1866.

3. FUCHSIA

L., Gen. Pl., ed. 5, 498 (1754)

Mostly shrubs, erect or climbing, less commonly trees or almost herbaceous. Leaves opposite, whorled or alternate, entire to serrulate. Stipules small, caducous. Flowers regular, normally pink, red or purple, rarely white or greenish. Sepals 4. Receptacle prolonged beyond the ovary into a " calyx-tube " or hypanthium, which is usually coloured and is caducous after flowering. Petals 4, rarely 0. Stamens 8, the 4 antepetalous ones shorter than the others. Stigma capitate or clavate, entire or lobed. Ovules pluriseriate in each cell of the 4-celled ovary. Fruit a few- to many-seeded berry. Seeds without hairs.

A predominantly American genus, with a few species in New Zealand and the Pacific, included here on account of a single apparently naturalized species. Many of the species have brilliantly coloured and attractively shaped, often drooping flowers and are therefore cultivated, and may be expected in East Africa. *F. boliviana* Carr., native from Jamaica and Guatemala to the Argentine, with more or less pubescent or puberulous stems and leaves, drooping red flowers whose hypanthium is 3–6 cm. long, and sepals 1–2 cm. long, is cultivated in Tanganyika, Njombe District, Aug. 1936, *C. J. McGregor* 14 !

F. arborescens *Sims* in Bot. Mag. t. 2620 (1826) ; Munz in Proc. Calif. Acad. Sci., ser. 4, 25 : 84 (1943). Type : a cultivated plant originating from Mexico

Evergreen shrub or small tree to 8 m. high, normally glabrous. Leaves opposite or in threes, elliptic to oblong-oblanceolate, mostly about 5–15 cm. long, and 2–5 cm. wide, acute or acuminate at both ends. Flowers red-purple, in erect corymbose many-flowered panicles. Hypanthium (see generic description) 3–8 mm. long, 1–3 mm. wide at top. Sepals linear to lanceolate, up to 9 mm. long and 2·5 mm. wide. Petals shorter than sepals. Berry subglobose, up to about 10 mm. thick, purplish (*fide* Munz, l.c.).

Tanganyika. Lushoto District : Kifungilo, 12 Aug. 1939, *Greenway* 5896 !
Distr. **T3** ; America, from Mexico to Panama, and perhaps native in Colombia
Hab. " Said to have been collected in open evergreen forest and to be fairly common "
(*Greenway* 5896) ; about 1220–1520 m.

The flowers of this species vary in size. The above-cited specimen has the hypanthium about 4·5–5·5 mm. long and the sepals about 5 mm. long and 0·75–1 mm. wide, thus coming under f. *tenuis* Munz in Proc. Calif. Acad. Sci., ser. 4, 25 ; 86 (1943), which appears to occur in America over the range of typical *F. arborescens*, but to be scarcer.

Grote 6545 !, in the East African Herbarium, was collected from a specimen planted at Amani. From there in the Eastern Usambaras, by distribution to other gardens and perhaps by birds eating the fleshy fruits, it no doubt reached Kifungilo in the West Usambaras. It remains to be seen how completely *F. arborescens* is naturalized in Tanganyika, but I consider that it merits inclusion in the flora. Botanists in the Usambaras should find out by what exact agency the seeds of *F. arborescens* are spread.

Mr. P. J. Greenway writes that *F. arborescens* is much cultivated in gardens in Kenya. For this reason further instances of its becoming naturalised should be looked for.

INDEX TO ONAGRACEAE